by Kathleen Powell

Printed in Mexico

ISBN-13: 978-0-15-362513-8

ISBN-10: 0-15-362513-9

2 3 4 5 6 7 8 9 10 126 10 09 08

Visit *The Learning Site!*
www.harcourtschool.com

Falling Satellites

In an old story, Chicken Little ran around warning all of her friends, "The sky is falling! The sky is falling!" You know that the sky isn't really falling—but did you know that the moon is? So are all of the artificial satellites that humans have launched into orbit around Earth. (A satellite is an object that orbits another object.) You don't need to go looking for someplace to hide, though, because the satellites won't fall to Earth. How can they fall without ever hitting the ground?

To understand how something can fall without hitting Earth, think about slowly throwing a baseball. When you throw, you push on the ball, causing it to move forward. If no one is there to catch the ball, it moves forward and curves downward before hitting the ground. Earth's gravity is pulling the ball down. If you throw harder, the forward force on the ball is greater, causing the ball to move farther outward as gravity pulls it down. The ball moves in an arc, or a curve. If you could throw the ball hard enough, the curve of its path would exactly match the curve of Earth's surface. Gravity would still cause the ball to fall, but the ball's forward motion would keep it from hitting the ground. (In practice, even if your arm were strong enough, you couldn't do this. Air resistance would slow the ball's motion.)

This is what happens to artificial satellites, too. When a satellite is launched, a force is applied to it. The satellite continues to move in the direction of that force because of inertia. Remember that inertia is the tendency of an object either to remain still or to remain in motion unless a force acts on it. In space there are almost no forces of friction or air resistance to slow the satellite's motion. However, the force of gravity still acts on the satellite. The satellite acts like the baseball you imagined. Gravity pulls it toward Earth, but inertia carries it in another direction. If the velocity of the satellite is just right, the satellite's path will exactly match the curve of Earth. The satellite is constantly falling without hitting Earth!

The same principle governs how the moon orbits Earth, how other planets' moons orbit those planets, and how the planets orbit the sun. One of Isaac Newton's great discoveries was that the same force—gravity—that causes objects to fall on Earth is also one of the factors that cause objects to orbit one another. This knowledge made it possible to understand and predict how planets and satellites move. Today scientists use this information to study the solar system. Scientists and engineers also use it to build and launch satellites that do everything from keeping an eye on the weather to retransmitting TV signals to giving astronauts a place to stay in space!

The harder you throw a ball, the farther it travels before hitting Earth.

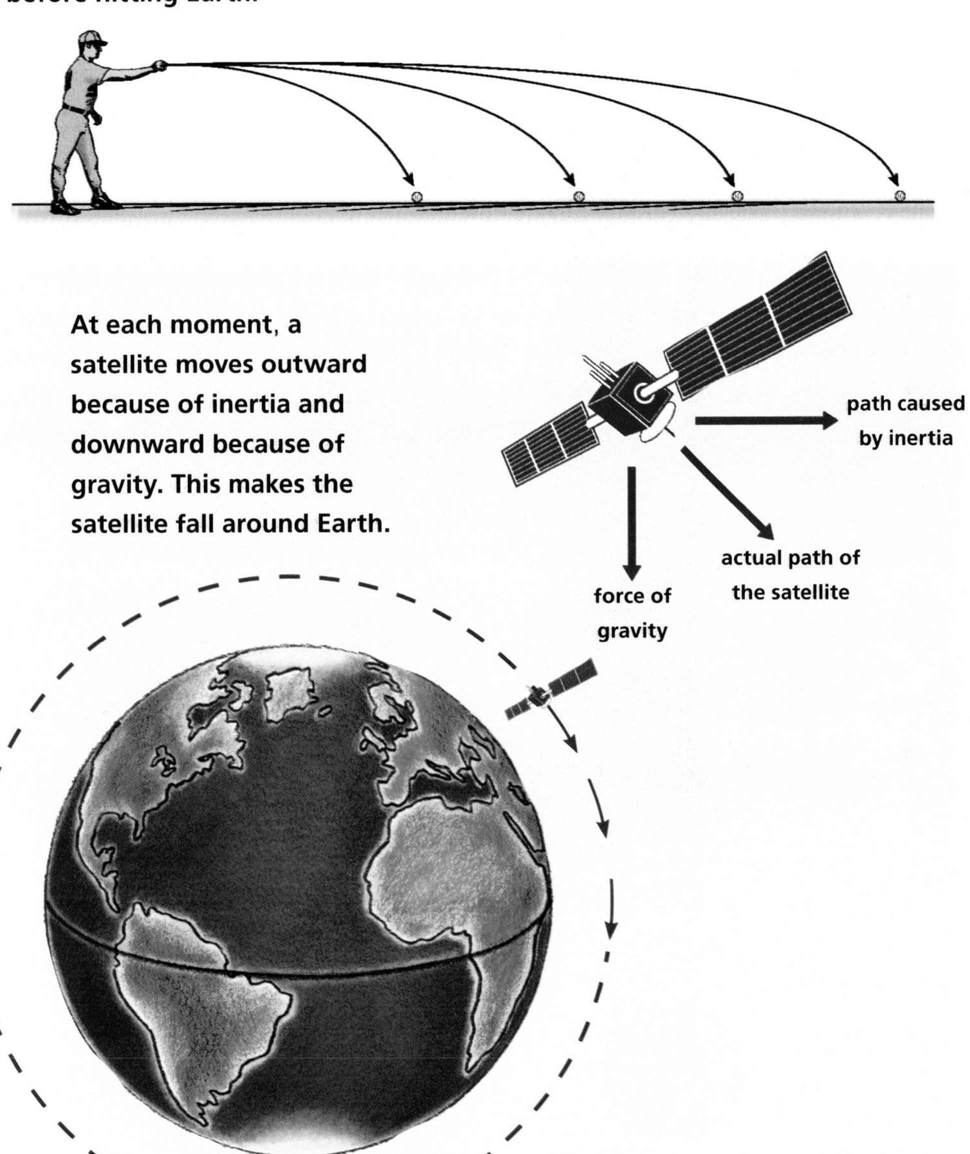

At each moment, a satellite moves outward because of inertia and downward because of gravity. This makes the satellite fall around Earth.

Orbits

Satellites can have many different types of orbits, depending on how fast they move and how far from Earth's surface they are. Scientists can decide whether a satellite will have an orbit that is low or one that is farther out from Earth's surface. They can make a satellite's orbit almost completely circular or more elliptical, like a circle that is stretched out. Scientists can also decide which parts of Earth's surface a satellite will pass over.

Satellites can have circular or elliptical orbits. Most are circular or nearly circular. Other satellites have more eccentric orbits. Eccentricity measures the difference between the closest and farthest points of the orbit. The closest point to Earth is called perigee, and the farthest is called apogee. In a noncircular orbit, the velocity and height of the satellite are constantly changing. When the satellite is at perigee, it moves faster than it does in the rest of its orbit. As it moves away from the planet, the satellite slows down until it reaches apogee. The satellite moves slowest at apogee and then begins to speed up again as it approaches Earth.

Incidentally, this pattern of motion in elliptical orbits is true for planets orbiting the sun as well. Planets move fastest when they are closest to the sun, and slowest when they are farthest from the sun. Planets are natural satellites of the sun!

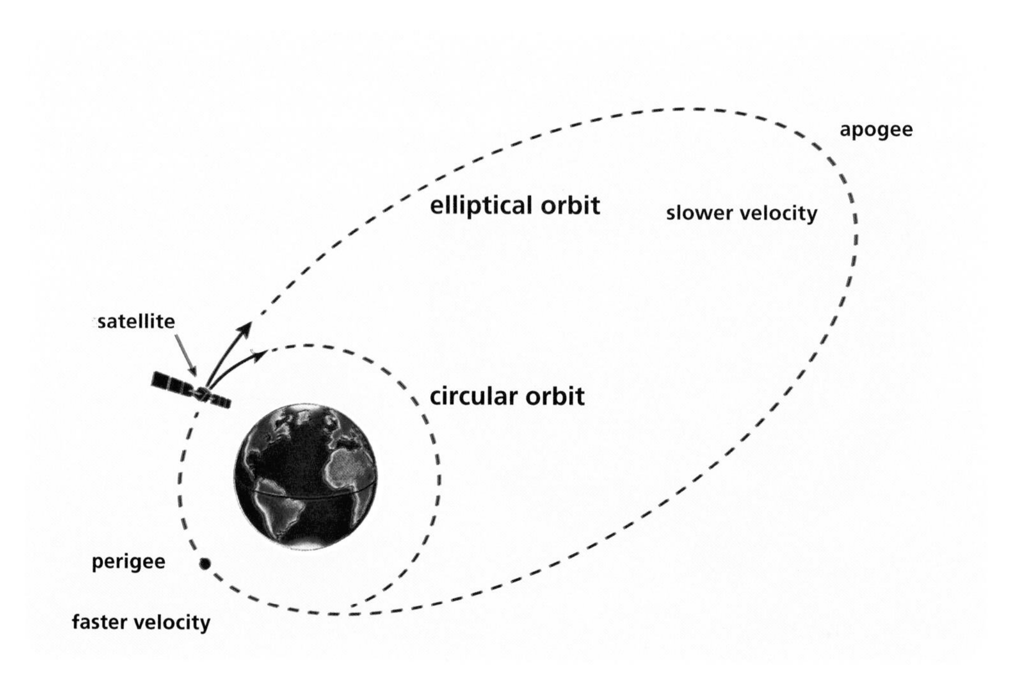

The rotational speed of a GEO satellite exactly matches the rotational speed of a spot on the equator. The spot rotates once in 24 hours. The satellite revolves around the Earth once in 24 hours.

The satellite's height above Earth must be 35,800 km (22,245 mi) and it must be in a circular orbit above the equator to maintain a geosynchronous orbit.

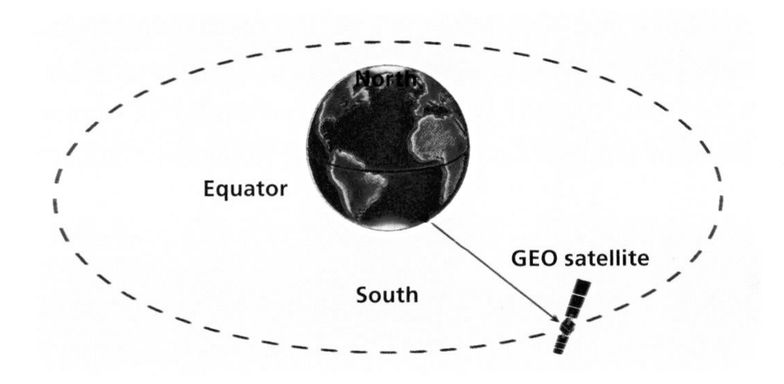

Geosynchronous Orbits

You probably know someone who uses a satellite dish to receive television signals. Have you ever wondered why these dishes can always point in the same direction, even though the satellite must be moving in orbit around Earth? It seems as if the satellite must stay in one place in the sky.

It may seem strange, but that is exactly what happens. The height of a satellite's orbit affects its speed. The force of gravity between two objects—such as a planet and a satellite—is related to the distance between them. A satellite closer to Earth must travel faster to maintain an orbit without having gravity pull it to Earth's surface. A satellite farther from Earth experiences a lower gravitational force. It can have a lower forward velocity and still stay in orbit. At a distance of 35,800 km (22,245 mi) above Earth's surface, an orbiting satellite moves in circular orbit at exactly the same speed as the rotational speed of a spot on Earth's equator. A satellite at this height that is moving in the same direction that Earth spins will always be above the same spot on Earth. This is why satellite dishes can always point in the same direction.

A satellite positioned in this way has a geosynchronous equatorial orbit (GEO). This name includes the prefix *geo-*, which means "Earth," and *synchronous,* which means "moving at the same speed." GEO satellites are used for communication, weather studies, and Earth observation.

Low Earth Orbits

The fastest-moving satellites have a low Earth orbit (LEO). Unlike geosynchronous satellites, LEO satellites can be placed in orbit anywhere around Earth. Their position above Earth is constantly changing because their rapid movement does not match Earth's rotational speed. The orbit of a LEO satellite may be about 300 km (186 mi) above Earth's surface, and the satellite takes about 90 minutes to complete one orbit. Because of their high speed and changing location, LEO satellites can be used for observing different parts of Earth. Low orbits are also used for astronomical satellites as well as for the space shuttle and the International Space Station because it is easier to place these satellites in low orbit than to put them into higher orbits. If you've ever seen photographs taken from the space shuttle or photographs of hurricanes taken from space, you know that low Earth orbits provide great views of Earth!

Medium Earth Orbits

Satellites with an orbital height between that of GEO satellites and LEO satellites are called medium Earth orbit (MEO) satellites. They orbit at a height of about 10,000 km (6,214 mi) above Earth's surface. Navigation satellites, which emit signals enabling people to determine their position on Earth's surface, as well as communication satellites, often use medium Earth orbits.

Polar Orbits

Most satellites travel from east to west, corresponding to Earth's rotation. Other satellites travel from north to south or from south to north, continually crossing over the equator. This is called a polar orbit because the satellites also cross over the North and South Poles. Because Earth spins from west to east, satellites traveling in a polar orbit have the capability to view the entire surface of the planet each day. This is very useful for mapping and weather observation. This makes polar orbits useful for satellites that monitor weather. These orbits are also used by scientists who monitor conditions on Earth, such as land use, oil spills in the ocean, or environmental destruction on land.

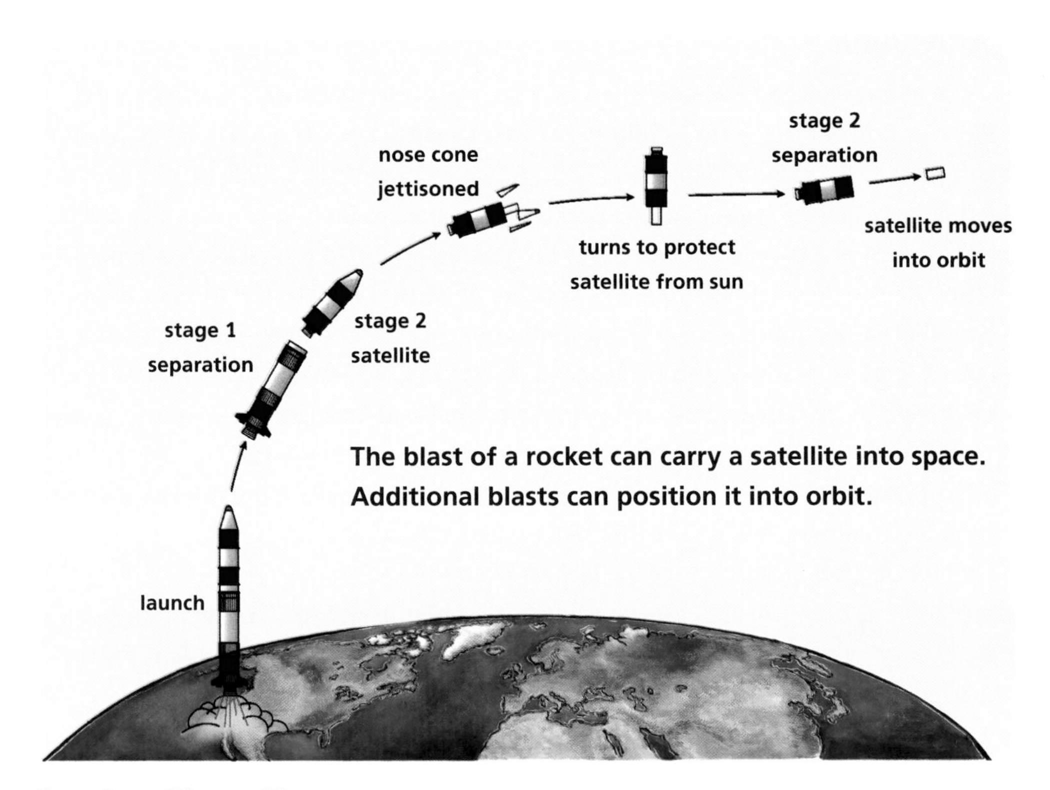

The blast of a rocket can carry a satellite into space. Additional blasts can position it into orbit.

Getting Them There

How do satellites get the momentum they need to get into orbit? Most satellites are launched into space aboard multistage rocket powered launch vehicles. The first stage for a typical 3 stage launch using a liquid fueled first stage lifts the rocket and satellite off Earth and through the atmosphere. The first stage falls away from the rocket after its fuel tank is empty. The second stage then ignites and lifts the satellite into a low Earth orbit. This stage then falls back to Earth as well. If necessary, a third rocket stage ignites and carries the satellite into a higher orbit until it reaches both the required height and required orbital injection velocity.

If the satellite will be in a geosynchronous orbit, it may need a series of short bursts from a rocket engine to get to the necessary altitude. First, a short blast of the rocket at perigee pushes the satellite from a circular orbit into an elliptical one. Then another blast when the satellite reaches apogee pushes it into a high circular orbit.

Satellites can also be carried into space in the cargo bay of the space shuttle. Astronauts on the shuttle make sure that the satellites are in good condition and then release them into orbit. Astronauts can also use the space shuttle to retrieve satellites that are no longer needed and fix those that are damaged. The shuttle's engines can even be used to boost satellites into higher orbits.

Space Junk

The moon and artificial satellites aren't the only objects in orbit around Earth. Millions of tiny objects, such as fragments that have broken off spacecraft, have been caught by Earth's gravitational force. Most of these fragments, called space junk, are less than a millimeter wide, but thousands of them are at least as large as a baseball. You may not think that a baseball could do much damage to a satellite or a space shuttle. But space junk may move at a velocity of 28,000 km (17,398 mi) per hour or more! Even tiny fragments of space junk can cause great damage. Space shuttle windows are frequently nicked by space junk. In 1983, a $50,000 space shuttle window had to be replaced when a paint chip smaller than the tip of a pencil struck it! In 1994, space junk hit a shuttle window, forming a pit about the size of a dime. In 1997, scientists had to use thrusters to move the Hubble Space Telescope away from an orbiting fragment of a rocket that had exploded several years before.

Scientists track thousands of the largest pieces of space junk. They use the information to keep satellites and spacecraft away. But tracking all of the fragments that are large enough to cause serious damage is impossible. When a space shuttle returns to Earth, it is inspected from top to bottom to determine damage from space junk. Each time, numerous pits are found where bits of debris in orbit around Earth have hit the shuttle.

Parts of a Satellite

Each satellite is unique. However, to stay in orbit and to survive the harsh conditions of space, a satellite must have certain parts. All satellites need a power source, a way to prevent wobbling, a body that protects the satellite against harm, and some way to communicate with people on Earth.

Satellites may have different types of energy sources. Batteries store energy, but they will eventually run out. You have probably seen many photographs of satellites that have winglike structures on their sides. These are solar panels that produce electricity by using energy from sunlight. Solar panels will not run out of energy, as a battery does. But solar panels work only when they are facing the sun, and the panels must be large enough to collect sufficient sunlight. Batteries can be used with solar panels and can be recharged when they run down. However, batteries are heavy, increasing the costs of lifting the satellite into space. Another possible source of energy is a small nuclear reactor; but if the satellite were damaged and fell to Earth, radioactive particles would cover the area.

Instruments inside the body of the satellite must be able to perform tasks such as determining the satellite's position and avoiding spins.

The body of the satellite must be able to withstand the radiation of space and collisions with space junk.

Solar panels provide power to the satellite.

Satellites must communicate with scientists on Earth by radio signals.

One problem that satellite designers must solve is wobbling. Special instruments in a satellite are used to detect wobbles and correct the satellite's movement. This type of correction is called attitude control. (When you're talking about a satellite, "attitude" means the direction that the satellite faces.) Wobbles can cause several problems for satellites. If a satellite uses energy from the sun as a energy source, a wobble can cause the satellite to turn away from the sun and have an outage. Wobbles can also hurt a satellite's ability to perform its job. For example, if the attitude of a satellite used for communication changes, the satellite will lose contact with people on the ground.

Satellites must have some way to avoid wobbling and maintain a steady orbit. One way to do this is to use small rockets called thrusters. The thrusters send out small bursts of gas in one direction, causing the satellite to move slightly in the other direction. Although thrusters are fast, they need fuel to work. After this fuel runs out, the satellite is no longer usable and may reenter Earth's atmosphere and disintegrate from drag and heat. Satellites may also control their position by using a wheel that spins rapidly. If the satellite begins to wobble, the wheel begins to spin, causing the satellite to correct its position. A third way to keep a satellite steady is with magnets that react to Earth's magnetic field.

The body of a satellite must be able to protect it from harm. Although scientists can steer the satellite away from large space junk, the outer layer of a satellite must be strong enough to withstand collisions with small particles. The outer layer must also protect the satellite from harmful radiation from the sun. The materials used in the satellite must keep the satellite from becoming overheated while it faces the sun or freezing while it faces away from the sun. The materials must also be light in weight, because the heavier a satellite is, the harder and more expensive it is to launch. It's a challenge to design a satellite body!

All satellites need some way of communicating with scientists on Earth. Whatever the main job of the satellite is, it must be able to send the data it records back to Earth. In addition, the satellite must be able to collect and send information about itself, such as temperature and position. Information that is collected by satellites is stored and analyzed by an onboard computer and then sent to Earth, usually using radio waves. Scientists on Earth can then send a signal back to the satellite, perhaps changing its position or changing the instructions for gathering data. Although tracking of a satellite is done on Earth, scientists must send this information to the satellite so that the onboard computer can determine when it is supposed to perform certain functions. Of course, the satellite's computer is also used to control all of its systems. The computer has instructions for what to do if it senses a problem with the satellite.

Types of Satellites

The first artificial satellite, *Sputnik* I, was launched into space in 1957. Since then, thousands of satellites have been launched, and today, over a thousand satellites are in orbit. They are used for communication, navigation, scientific research, and military applications, as well as for monitoring weather conditions.

You may be most familiar with communications satellites. Every time you watch television or use a cell phone, you probably use a satellite link. The signal that allows you to watch television may have started a long distance away. The television station transmits the signal to a satellite, and the satellite relays the signal to a ground station near you. This station may then send the signal along a ground cable to your house. The signals to and from cell phones and pagers can be sent in a similar way.

The satellites used for communications are often placed in geosynchronous orbits. This allows a satellite to stay in constant contact with a ground station. Other systems use dozens of LEO satellites placed over Earth's suface. This allows people to communicate even in polar regions and other areas not covered by geosynchronous satellites.

Remote Sensing Satellites

If you wanted to find out whether a forest had been affected by pollution or what type of mineral resources an area has, what would you do? You could go visit the area, but then you could see only a little of the area at once. What if you could see it from space? You could see the whole area at once. Remote sensing satellites let people do just that. Remote sensing satellites can be in either geosynchronous orbit or low Earth orbit. Geosynchronous orbits allow the satellites to keep one viewing area at all times. Low Earth orbits allow the satellites to obtain clear, up-close images of many areas.

Remote sensing is often used to monitor Earth's weather and to monitor geologic and environmental conditions on Earth. One of the most familiar uses of remote sensing satellites is for weather maps and forecasts. Maybe you have seen or heard a weather forecaster talking about what is shown on satellite maps. These maps are obtained by using weather satellites. Some satellites have sensors that can determine the amount of water vapor in different parts of the atmosphere. Satellites can also monitor the movement of warm and cold air masses. When tornadoes or hurricanes threaten populated areas, weather satellites can monitor the movement of the storm systems, enabling people to avoid dangerous areas. Scientists can also use satellite data to study weather patterns over time and identify water and wind currents.

Remote sensing satellites have many uses in studying and protecting the land and water.

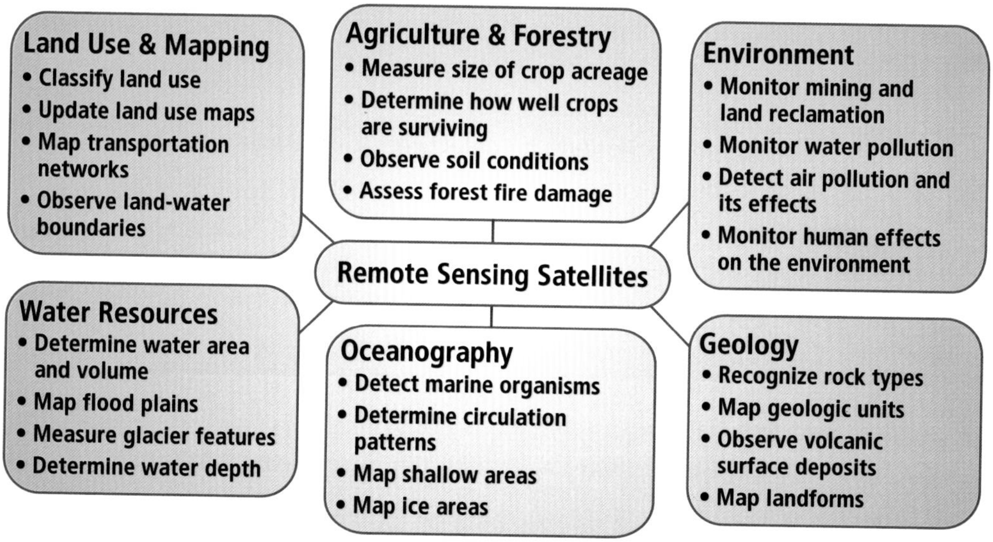

Remote Sensing Satellites

Land Use & Mapping
- Classify land use
- Update land use maps
- Map transportation networks
- Observe land-water boundaries

Agriculture & Forestry
- Measure size of crop acreage
- Determine how well crops are surviving
- Observe soil conditions
- Assess forest fire damage

Environment
- Monitor mining and land reclamation
- Monitor water pollution
- Detect air pollution and its effects
- Monitor human effects on the environment

Water Resources
- Determine water area and volume
- Map flood plains
- Measure glacier features
- Determine water depth

Oceanography
- Detect marine organisms
- Determine circulation patterns
- Map shallow areas
- Map ice areas

Geology
- Recognize rock types
- Map geologic units
- Observe volcanic surface deposits
- Map landforms

Remote sensing satellites used to monitor geologic and environmental conditions have a large number of uses. When a volcano erupts, satellites can record the movement of ash clouds, allowing scientists to warn people if the clouds move toward a populated area. Satellites can also monitor forest fires, showing how they may spread and what might be done to prevent them. Tracking smoke and ash from fires and volcanoes helps scientists study wind patterns.

Satellites can take special photographs that allow scientists to map various geologic features and monitor pollution of land and water. Scientists can use computers to analyze these photographs. Images obtained by remote sensing satellites help scientists learn about landforms and mineral resources. They can also be used to help make accurate maps.

Navigation Satellites

If you were lost somewhere in a city, you could look on a map and use landmarks to determine your location. But what if there were no landmarks? What if you were in the middle of the ocean, or deep in a forest? Under these circumstances, maps wouldn't help you very much. In places like these, you can use navigation satellites to locate your position. *Navigation* means "finding your way." Navigation satellites use radio signals to help you determine your position on Earth's surface.

The Global Positioning System (GPS) is a network of 24 navigation satellites placed in orbit at certain places above Earth. These satellites enable people with the correct type of receiver to pinpoint their location on Earth's surface. Although the satellites were originally used only by the military, they are now widely used by hikers, hunters, boaters, and any other people who want to know precisely where they are. Some cars now have a GPS tracking system installed. The system works with a computer to show the car's location on a map.

A GPS receiver determines position by detecting radio signals from four GPS satellites. Each signal contains information about the exact time it was sent from the satellite. It takes time for the signal to travel from the satellite to the receiver. A receiver can determine its distance from the satellite by using the difference between the time that the signal was sent and the time it was received. The signal also contains information about the satellite's location. By finding its distance from three of these satellites, the receiver can identify exactly where on Earth's surface it is. This process is called triangulation. The diagram on page 13 shows how triangulation works. The signal from a fourth satellite enables the receiver to determine its height above sea level.

By measuring its distance from three satellites, a GPS receiver can narrow its position to one of two places. Usually one of the places is clearly too far away. If necessary, a fourth satellite can eliminate the wrong choice. A fourth satellite can also identify how high above sea level the receiver is.

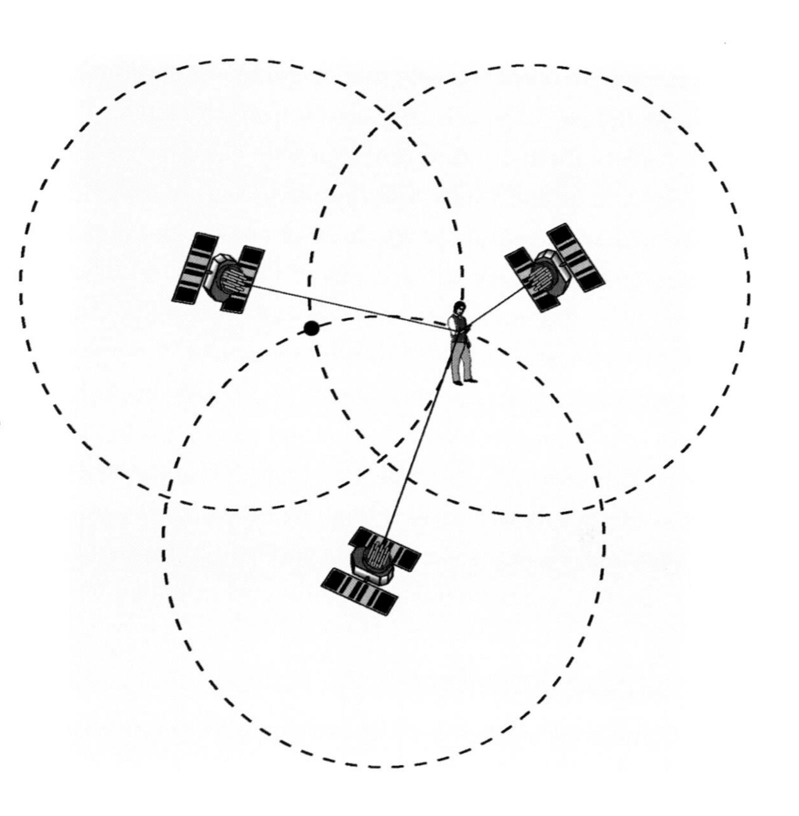

Astronomy Satellites

Have you ever looked up at the night sky and seen stars twinkling in the distance? These stars are a beautiful sight, but if you are an astronomer, twinkling isn't a good thing. A star seems to twinkle because the pinpoint of light is disturbed as it passes through Earth's atmosphere. You see a similar effect if you stand in a swimming pool and look down at your feet. Light is disturbed by the sloshing water, so your feet seem to move back and forth when they really aren't moving at all. In the same way, when light travels through air, the light is deflected in several directions. Light from stars, then, seems to move slightly back and forth. When astronomers look through telescopes at distant galaxies and astronomical phenomena, this twinkling can disturb the images enough to prevent the astronomers from making useful studies of stars.

Viewing the sky from high altitudes reduces the twinkling, but there is no way on Earth to get rid of it completely. To avoid the twinkling, you need to take telescopes above Earth's atmosphere. Astronomy satellites are telescopes orbiting Earth. These telescopes can obtain clearer images than land-based telescopes. They can also see very faint objects because they are far from city lights.

Many of these orbiting telescopes are designed so that they receive different wavelengths of light, such as X rays, microwaves, infrared light, and gamma rays. The human eye cannot see this form of energy, but many objects in space give off energy at these wavelengths. Some of these wavelengths are absorbed by Earth's atmosphere, too. Images from these sources provide scientists with more information than they can get from Earth-bound telescopes. Computers can add color to images in these wavelengths so that people can see the details in them.

One of the most famous astronomy satellites is the Hubble Space Telescope. Hubble is the largest telescope ever built as a space observatory. It was launched in 1990 from the space shuttle into a low Earth orbit so that it circles Earth once every 95 minutes. The many instruments and imaging systems on Hubble allow scientists to see some of the most detailed pictures ever obtained of distant space objects.

Occupied Satellites

Some artificial satellites can carry people. The space shuttle is like a temporary satellite. While it is in orbit, it acts like a satellite. When it is time for the space shuttle to land, the astronauts use its engines to change its inertia so that it falls toward Earth in a controlled way. Other artificial satellites are designed to be homes for astronauts for long periods. These are often called space stations. The International Space Station (ISS), for example, is designed to be occupied permanently by crews of astronauts. The ISS was carried into space in parts by space shuttles and assembled by astronauts.

You read earlier that the ISS has a low Earth orbit. This makes it easy for the space shuttle and other spacecraft to reach the station. Low Earth orbit can present problems, though. Earth's atmosphere causes a force called drag, or air resistance. This force slows the motion of the satellite, which changes the arc of its orbit. Remember the baseball? When it had lower velocity, it fell to Earth sooner. The same thing can happen to a satellite. As the drag slows the satellite, it falls closer to Earth and the force of gravity becomes stronger. The satellite starts falling even faster, and if nothing changes, the satellite will crash to Earth. To prevent this from happening to the International Space Station, astronauts use fuel from the space shuttle to push

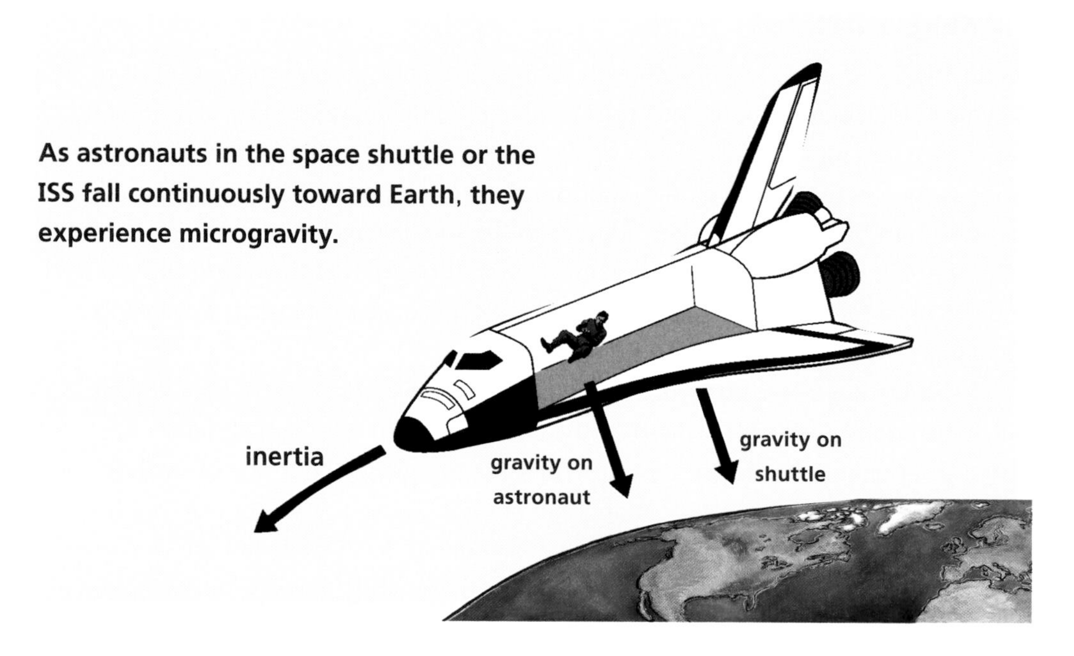

As astronauts in the space shuttle or the ISS fall continuously toward Earth, they experience microgravity.

the station into a higher orbit. This has to be done repeatedly because the station falls about 9.65 km (6 mi) each month!

Have you ever seen photographs of astronauts in the space shuttle or ISS? If so, you may have seen them floating in midair! People often talk about space as a "zero-gravity" environment. This does not mean that there is no force of gravity on the astronauts, though. You know that an orbiting satellite is still affected by Earth's gravity. Without gravity, the satellite would fly off into space.

What people really mean when they say that astronauts live in zero gravity is that they are in a constant state of what's called free fall. Free fall is what happens to an object when the only force acting on it is gravity. Suppose that you jumped into a very deep hole. Gravity would cause you to keep falling for a long time. Now suppose that you were holding an apple and let go of it as you fell. The apple would be falling at the same rate, so it would seem to hover next to you. (In real life, air resistance would change how you and the apple fell, but let's ignore that for the moment.) You and the apple would both seem to be floating!

This is what happens to astronauts. While they are on the ISS (or in the space shuttle), they are constantly falling. The satellite and everything inside it are falling at the same rate. There's no air resistance, so gravity is the main force acting on the satellite and the astronauts. This creates what is called a microgravity environment. Microgravity means that the effect of gravity is very small.

Viewing Satellites

In 1957, people watched eagerly as *Sputnik*, the first satellite, crossed the sky. Radio and television stations broadcast the beeping of its signals. Today, it's much easier than you might think to spot satellites! All you have to do is know where and when to look in the night sky. Although satellites don't produce their own light, you can see them because they reflect light from the sun, just as the moon does. In fact, you may have seen satellites in the night sky and thought that they were airplanes. You can tell a satellite from an airplane, however, because airplanes have blinking red lights.

A satellite looks like a star moving across the sky. Because satellites are visible only when the sun shines on them, you are most likely to see them about an hour after sunset or an hour before sunrise. This is the time when your side of Earth is in darkness but when the satellites are high enough above Earth's surface that the sun still shines on them. The speed of a satellite depends on its height above Earth. A typical satellite may take about 15 minutes to move across the sky, but some can move across in even less time.

You can increase your chances of seeing a satellite by preparing ahead of time. The first thing you need to do is plan a viewing place. Whether you live in a city or in the country, plan to go to a spot that has as little light as possible. You will also need to find a satellite that you will be able to see. The satellites must be bright and visible during the time you plan to view them. You can find details on the Internet about viewing satellites in your area. Write down the path, such as north to south, that each satellite will take, and the time when it will be visible. It's also helpful to determine when the satellite will pass close to certain constellations or stars. If you aren't familiar with the constellations, study them ahead of time and look for them in the night sky.

The satellites you will see will be in low Earth orbits. Although geosynchronous satellites stay in one place, they are too high to see without expensive telescopes. A space shuttle in orbit is a good satellite to view because it is so large. The International Space Station is also very visible because of its size. The National Aeronautics and Space Administration (NASA) posts information on the Internet about the best times and locations to view both space shuttles and the International Space Station.

GPS satellites are also easy to spot. These communication satellites have antennas that are like highly polished mirrors. When sunlight reflects off these satellites, a bright flash of light is visible. You can also find information on the Internet about when these flashes will be visible in your area. Keep your eyes on the sky. You never know what you might see!